Medical Exam.
Cambodia

Medicine Dentistry Pharmacy
Nursing Midwifery

SENKAWA TOMOO

2 MEDICAL EXAM. CAMBODIA

Rapidly sought delivery teacher, nursing teacher

4 MEDICAL EXAM. CAMBODIA

Also, by SENKAWA TOMOO in Japanese as paper books

- Passion today's Education Cambodia/ISBN 9784990975289
- KANBOJIA no AOJASHIN/ISBN 9784990293123
- Fantasy? Cambodia northern train/ISBN 9784990557041
- Amazing! wheels evolution in Cambodia/ISBN 9784990975203
- Four god's faces ! Angkor Wat Cambodia/**ISBN 9784990975265**
- inside PHNOMPENH/ISBN 9784990293109
- Landmine field's bumpy road in Cambodia/ISBN 9784991090332
- Fine! Cambodia southern train/ISBN 9784990557058
- Remote ancient city ruins Cambodia/ISBN 9784990975272
- Capital KINDERGERTON boom in Cambodia/ISBN 9784991090318
- MIZU KANBOJIA/ISBN 9784990293178
- Yama to KUSAKI KANBOJIA/ISBN 9784990293185
- Amazing! dancing cranes in capital! Cambodia/
ISBN 9784990557096
- Passion! clickety clack trial run northern train Cambodia/
ISBN 9784990975241
- Capital car society has come! in Cambodia/ISBN 9784991090349
  etc.
  all from AJIA Kobo HENSHUSHITSU

Copyright 2021 by Egashira Shoichi

All rights reserved.

Published by AJIA Kobo HENSHUSHITSU

Originally published in Japanese
Medical course exam in Cambodia

By AJIA Kobo HENSHUSHITSU, Tokyo, Japan 2020 Mar.

Medical exam Cambodia
As e-book of English

Translated published in English

Manufactured in Japan

First English edition

By AJIA Kobo HENSHUSHITSU, Tokyo, Japan 2021 Dec.

Author career

SENKAWA TOMOO (real-name Egashira Shoichi)

I give up entering, acquiring the permission of the studying abroad of University of Berlin/Berlin Humboldt University at the united Germany previous year, too.

My travel of Africa/South America/etc. runs through in front of my bumpy mind just after that.

I was born in Fukuoka Prefecture, Japan.

GRADUATION of Tokyo Metropolitan University.

* In this book,

Photos and Drawings are included

about Improvised sketches, English

translations and so on.

All from   SENKAWA TOMOO

/Egashira Shoichi

with using some kinds of PC soft.

8 MEDICAL EXAM. CAMBODIA

Delivery teacher who is holding a baby in a population explosion

This book is the contents about the Cambodian medical treatment.

A doctor system university is only in PHNOM PENH for several school. 1 school was the only 1990's first half.

An entrance examination pass magnification in those days was 30 times to 40 times.

The competition that can't be thought of absolutely existed.

The state of the medical scene should be permitted to be told originally but it was decided to concern the medical exam of that runup.

It was decided, after that, for a quarter of century to pass about how to become and to see that state again.

It is the thought that touched the wind of the new medical world today, too.

10 MEDICAL EXAM. CAMBODIA

prologue

In Cambodia, there is an influence from various countries only by being involved in medical care.
For example, it is new news that a Japanese hospital has been opened from Japan.
When I actually went to the hospital, many Cambodian people visited the hospital, and the presence of this hospital was large.
It seems to be fulfilling so that it is thought that the patient of a serious illness in Cambodia might not have to be hospitalized abroad if there is such a splendid hospital.
In Cambodia, there are people who are forced to be hospitalized in Singapore, Thailand, etc. overseas.
Therefore, the hospital which imports the hospital from the rich country directly as it is reflects like the luxurious one.
In addition, doctors from Europe and the U.S. are active in Cambodia.

And by chance meeting face-to-face, it seems that they are really supporting the local people from the feet.
The reason why I ended up making this booklet was because I used to be involved in medical examinations in Cambodia.
It happened to touch people who experienced the examination in the civil war age by chance.
There was such a thing, and I tried to touch the current state of today's examination again.

12 MEDICAL EXAM. CAMBODIA

table of contents

Prologue・・・10

Chapter 1 From the former examination situation
・・・16
Chapter 2 Medical and Dental Pharmaceutical Universities Located Only in PHNOMPENH
・・・31
Chapter 3 People involved in medical care and people studying medical care
・・・42
Chapter 4 Female Doctors Studying Abroad under the Scholarship Program・・・55
Chapter 5 Preparatory Schools Attended by Active Medical, Dental and Pharmaceutical Sciences・・・68
Chapter 6 Situation of Foreign Languages in the Faculty of Medicine, Pharmacy and Dentistry・・・85
Chapter 7 Nurses and Midwifery Schools Born Here and There・・・94

Chapter 8 Break time・・・１０８

Chapter 9 The Entrance Examination Ratio of the
         Faculty of Medicine, Dentistry and
         Pharmaceutical Sciences・・・１２３

Chapter 10 In 1994, the test magnification of
          the School of Medicine was even more
          accelerated・・・１３７
Chapter 11 It is hard to enter and graduate from
          medical college, and it is hard
          again after entering medical school
          ・・・１４９
Chapter 12 Rehabilitation Welfare Facilities
          ・・・１６２
Epilogue・・・１７３

14 MEDICAL EXAM. CAMBODIA

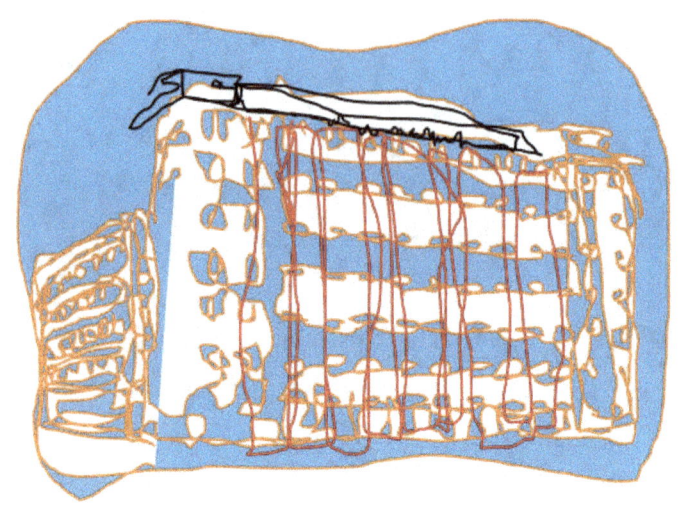

It is the university which a medical department system university existed in at the 1990's first half.

It was the water, thing even what's doing every day besides really no matter how by this you may be spending every day on then in capital Phnom PENH.

There weren't the water and food to that much.

The water in a lodging is the brown water.

However, heat sunlight drifted, and a body calmed down by being bathed in the water from the daytime.

The magnification of a university so is 30 times and 40 times inside.

Probably, will the person who is reading such a word understand?

It seems that it is surprisingly an incomprehensible thing when being in an abundant country.

Chapter 1

From the former examination situation

It is represented as

"FACULTE DE PHARMACIE".

18 MEDICAL EXAM. CAMBODIA

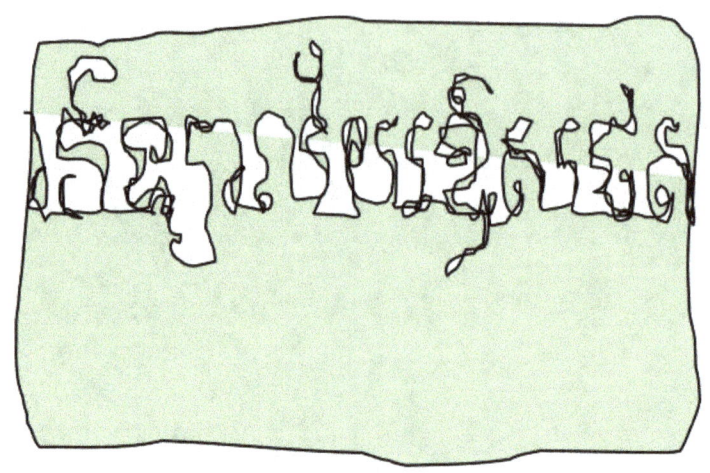

「 FACULTE DE PHARMACIE」 is written

From previous page

It seems that it is surprisingly an incomprehensible thing when being in a country with the abundant loss of all things by a civil war.

There aren't both the drinking water and one to eat when the daily life is lost and also there isn't the education or medical care or anything.

There isn't the electricity or anything.

If there isn't an amusement, either, there isn't really.

There may not be the people in the neighborhood, too.

20 MEDICAL EXAM. CAMBODIA

Ceiling in a hospital and state of a passage

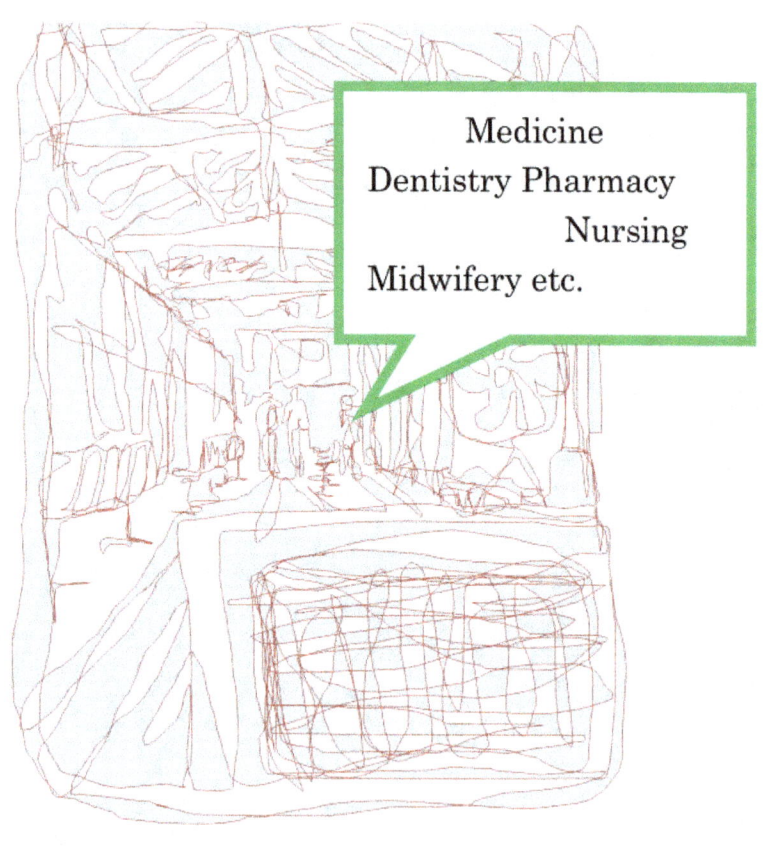

Medicine
Dentistry Pharmacy
Nursing
Midwifery etc.

From previous page

It is a state by a hospital by a local city.
Window in a ceiling and reception desk on both sides.
To it, in front, a doctor and patient sit on a chair.
In an improvised sketch, it is incomprehensible but is anyway a state inside a hospital.
The person who is in the center that becomes the condition that we understand with what the electric fan that is on the right is the center is a doctor and nursing teacher.

22 MEDICAL EXAM. CAMBODIA

From previous page

It runs in capital Phnom PENH, ambulance.
Of course, it is the best that there isn't such a thing
But however, the various accidents occur every day in Phnom PENH.
A traffic accident is what is occurring in the increase of a car in the various places.
It seems that it becomes gradually as it is possible to be dispatched immediately at such time.
1990's didn't have an ambulance.
Of course, we think that there were very much few car crashes, too.

24 MEDICAL EXAM. CAMBODIA

"Scenery of medical examination"

I had touched the university examination rather than the site of the medical treatment.
In the early 1990s, during the civil war, I was involved by chance.
So, I will hear a lot of things.
This was that the medical examination was heavy in the state of the stage before the medical treatment site.
I had met the person who took the examination of medical school, the school of dentistry, and the pharmacy under the civil war and people who took the examination.
I did not think why there was such a terrific entrance examination situation under such a civil war to the reality.
The reason is that the event which enters my view, and it does not seem to be a reality was happening.
Nevertheless, paradoxically, this was happening.
"What can't happen normally in any field is happening in Cambodia."

It was.
Then, it was that I had the skin and did not notice it.
"I only know the normal world of everyday life in a rich country."
"For that reason, I don't know what extraordinary things are."
It was.
It was that I finally noticed the locale.
"The word 'anomaly' is too routinely used. But I'm not aware of the degree of 'anomalies'."
It is.
The civil war killed many people.
Among them, many doctors are included, and there is no one to save those who suffer from the disease. Moreover, people related to the medical treatment were almost lost, too.
There was no educator and only the engineer at all.
At that time, Cambodia lost the people who supported all social infrastructures.

There is a shortage of doctors, dentists, etc.

26 MEDICAL EXAM. CAMBODIA

Meanwhile, in PHNOM PENH, the capital of Cambodia, in the early 1990s, I know the reality of the examination.
That makes it like this.
"How will medical care proceed in the future?"
The question, to me, was spreading.
Then, in front of me, a quarter of a century passed.
From around 2015 to around 2020, I was beginning to see a new progress in medical matters.
I saw and heard today's situation through the examination related to today's medical treatment again.
On the way, in the 1990s, I also saw a mixture of medical, dental, and pharmaceutical departments and other aspects of today's state.
It used to be like this.
"There is an urgent need due to the shortage of medical professionals such as doctors and dentists."
And, I had heard from the student of the medical school.

In addition, in a simple hospital in the city, drips were performed on patients lying on the bed. That came into my sight without permission from along the road.
It became very little today that the inside of the hospital on such a road side was easily seen to me.
The building of the hospital has improved.
There is such a thing.
Too much heat radiation drifts. Also, sweat flows from my forehead. In a building without air conditioning, it looks like this.
"Is it already okay before the patient's body is treated?"
I became, too, to feelings.

What the "snake sign" informs

In addition, in the early 1990s, it was like this in the city.
"I drew a snake"
There were too many signs, and I was able to see the pattern.

28 MEDICAL EXAM. CAMBODIA

Because the number of doctors and dentists was extremely low,
"It was a society in which we had to rely on medicine."
It was.
Therefore, it was this way in capital Phnom Penh in Cambodia.
"Why are there so many signs related to medicine?"
It was much later that I realized.
Anyway, there is a pharmacy with a signboard along an open, simple hospital and the road in the hot air. So, this is the last line where people can take care of medical facilities and medical institutions. This seemed to me, like a choppy, last-ended countermeasure, of the people.
Then, a lot of patient's people existed from a lot of signboards, and the appearance which could not help relying on the medicine seemed to have been informed to me indirectly.
Because it is this way, the drugstore is just like this.

"How do you deal with many customers?"
And, I was worried, and there was a time.
"I don't have enough people, I don't have enough medicine."
It was.
As for pharmacies, many years have passed.
Then, I did not notice that such circumstances became the one which extended more widely today.
With the population explosion, it was continuing.
"The need for nurses and midwives was also urgent at the time."
I did not notice, too, at once to that.
It was not only the field of the medical treatment that I did not notice.
This is what it is.
I did not understand how much serious entrance examination had been waited in the faculty related to the medical treatment of the university study.
I meet the Cambodian people. Then, I was able to know the realities for the first time by me hearing the content directly.

30 MEDICAL EXAM. CAMBODIA

During the last quarter century, medical trials have undergone a change.
Around 2008, the improvement was observed.
After the Grade 12 test to prove high school graduation, the time for the exams for the other entrance exams held by each university was shifted. In the medical examination in Cambodia, the reduction of the examination subject was given, too.
It was a good turn for the examinees.
However, the number of people related to the medical treatment is still not enough.

Chapter 2

Medical and Dental Pharmaceutical Universities Located Only in PHNOM PENH

32 MEDICAL EXAM. CAMBODIA

Building of a doctor system university.

From previous page

A university about the medical treatment is only in Cambodian capital Phnom PENH.
Moreover, they are only several school.
It was only 1 school per 1990's.
A quarter of century passed and became 3 school today.
If there are both today and person who study abroad overseas for the medicine, it looks for a hospital for the hospitalization in the nearby various countries.

34 MEDICAL EXAM. CAMBODIA

In Phnom PENH, a medical department dentistry pharmacology department system university only about 3 school.

※ It was possible to hear the situation of an exam in the 1990's first half, building from a department of dentistry.

A thing such as this is occurring absolutely by the civil war times wasn't believed.

The university that has a medical school, department of dentistry, school of pharmaceutical sciences and so on fine increased today in a faculty of liberal arts, university of a science faculty.

A doctor system university became 3 school when including a university at present.

However, even if there aren't a medical school, department of dentistry and pharmacy in a university, the school of nursing science really is in a science faculty, faculty of liberal arts department.

There is the department of veterinary science at the same time in a university of agriculture, too.

There is the university that is applying for the approval as a medical college in additionally, recent years in a country for more than one school, and a doctor system university or university that have a medical system department may fine increase.

36 MEDICAL EXAM. CAMBODIA

a medical, dental and pharmaceutical university located only in Phnom Penh

From those related to hospitals,
I got the opportunity to hear about the school to become a doctor or dentist.
"This is not in rural areas."
"This is limited to medical and dental universities in Phnom Penh."
"Until around 2007, there was an entrance examination before graduating from high school."
"But at that time, more people failed the previous year of the exam."
"So there were circumstances where I couldn't prepare for the entrance exam."
"At that time, there was a discussion of changing the examination date."
As a result, it is like this.
"After 2007, the examination period has changed since 2008."
"The entrance examination was moved in November after the national common high school graduation exam to be held in August."

"This allows students to prepare for the entrance exam between three and four months of graduation."
"During this period, students were able to study."
And explanation.
Now, students can take the examination after studying enough before taking the examination. It is said that the system of the examination changed to the system.
The examination subject is like this, too.
"Until 2007, there were seven subjects."
"Since 2008, there have been three subjects, including mathematics, biology, and chemistry. So, this is less."
And explanation. Until then, it's like this.
"English, society, national language, French, etc. were added."
"Until 2,000 years ago in the entrance examination, the magnification was over10 times higher than the entrance examination."
"Even before 2008, there was a shortage of doctors in Cambodia."

38 MEDICAL EXAM. CAMBODIA

Therefore, the government was moving to increase the number of students in the medical school.
"The system of the examination was changing in the direction of reducing the number of subjects in the examination."
The word continues.
"It's hard to get today's magnification accurately."
"But for example, the total number of entrance examination students in the 200 medical school quotas is 1,000."
"Then, at almost 5 times the magnification, students can take the entrance examination."
And explanation. Still, the examination of the medical school is difficult, and everything is the same.
"I graduated in August and I have a Grade 12 exam in mid-August."
"After that, I will have the exam in November."
"But I may fail the Exam in November."
"In that case, students will go to preparatory school until the next year's exam."
Therefore

"Students intensively study exams in all subjects, including mathematics, biology, chemistry, and physics."
It is.
This prep school looks like this.
"Not in rural areas, but in Phnom Penh."
There is a medical, dental and pharmaceutical university in Phnom Penh.
Therefore, the preparatory school and the cram school seem to be concentrated in Phnom Penh.
Of course, other difficult universities are also concentrated.

40 MEDICAL EXAM. CAMBODIA

Medical expert, medical training leader in the foreign countries that came to Cambodia, Cambodian medical student and teacher of a nursing teacher/delivery teacher who run around everywhere who came from near a Vietnamese and Cambodian border.

※ The people come to Cambodia from the various countries.

The medical world, too, is the same in that thing.

Person's Cambodian activity, too, is increasing today in the medical world.

We are only surprised at the whole country palmer of the person of a nursing teacher/delivery teacher, too.

The state of medical system college student's learning, too, has the power.

Chapter 3

People involved in medical care and people studying medical care

Male nurse

From previous page

A person about nursing, delivery sees the Cambodian country the figure whose whole country is moved for that lecture around the time when it passes 2000's in the Cambodian people if there were the people in the foreign countries that came to Cambodia, too.

Also, student of a medical college, too, is very strenuous in an examination.

He is medical college student's person who becomes strenuous in the schoolwork in a medical college today.

Founding official of medical school, European American surgeon

I happen to meet a former American surgeon teacher,
I am the person who met when I am moving in Cambodia.
He participated in the creation of a medical university in Phnom Penh, the capital of Cambodia.
The person is a person who is doing medical training and medical guidance.
This person is one of those who found the founding person of the medical university only by counting in the capital Phnom Penh in Cambodia. He is an American doctor born in Europe.
I became a doctor in my home country.
After that, I go to America.
"So, I start out as a surgeon practitioner."
"After that, I will continue to be a surgeon at medical institutions, including New York on the East Coast."

46 MEDICAL EXAM. CAMBODIA

"After that, before I know it, I belong to the American medical corps, and my heart leans overseas."

"Then I was supposed to be involved in medical care, such as volunteering."

This is a surgeon who changed from the medical practitioner to the medical treatment of the volunteer so to speak.

"At university, I leave a lot to Cambodian medical professionals today."

That's how powerful Cambodian medical professionals are doing.

a university student attending a medical school

I was able to meet people of the medical school in capital Phnom Penh.

The reason is that it was possible to visit the office of the medical school to me.

Even though it was a heat radiation, I was able to go through the gate of one of the medical universities that only have three schools in Cambodia.

Here, the faculty of medicine, the school of dentistry, the pharmacy, the faculty of nursing, etc. were made.
It is a university where the name of the university is written in the building which entered a little side from the main street.
I go in with the student.
Then, there is an office, and the pamphlet is arranged.
Time seems to be the student's holiday time. It's a square that might be in detail.
However, there are students who sit in the chair in the plaza of the university, and study.
Then, there is a person who is eating, too.
There are a lot of students.
At that time, I was able to talk with some students.
By chance, I was able to talk to students from the School of Dentistry and pharmaceutical students.
Several others joined us.
One student said,

48 MEDICAL EXAM. CAMBODIA

"Today, I have an exam from12:30, so I come to school early to study."
At that time, a friend joined me in the story.
The first student is like this.
"I'm from PREAH VIHEAR."
"At this university, 80%of students have passed the national exam."
Other students are like this.
"We are from Phnom Penh and go to this university."
Then, the content inclined to the story of the national examination.
It also extended to the story of the difficulty of the examination in the university.
The student is registered in the medical school very. However, the talk has jumped out about the talk of the preparatory school for the examination in the university.
I went to the prep school to enter college.
"I have something to go to today as well."
In addition, today is an examination day, and it is like this.
"In an hour, we'll have an exam."

The first student is from a rural area, so it's like this in Phnom Penh.
"My cousin, who is four years older, and I have the same boarding house."
In Phnom Penh, in an apartment near the university, it looks like this.
"I'm renting a $100-a-month apartment."
"We're giving each other money for$50."
And that.
At this time, it is an examination of the specialized subject of each grade. Then, students' heat overflowed anyway by the appearance which worked hard at study.

*The national examination did not come out deeply here.
The pass rate here came out by chance.
But other than that, I came to hear. Later, I found out, that the national examination had never been so long since it began in the early 2010s.

50 MEDICAL EXAM. CAMBODIA

The student had been given doctor's qualification as it was if the student graduated from medical school till then.
However, the student must pass the examination of each subject of each grade at the university. On top of that, the last new national examination has begun in Cambodia.

Busy going everywhere, nurse and midwife teacher, teach hundreds!

The teacher is like this.
"I grew up on the Cambodian side near the border between Vietnam and Cambodia."
"My parents are a family of Chinese fathers and Vietnamese mothers."
"I experienced the Pol Pot period from 1975 to 1978."
"I was engaged in agricultural work in the northeastern part of Cambodia."
Before I went to Tohoku, it was like this.
"At first, I lived in Phnom Penh and was17 years old."

"At that time, I spent my time in Phnom Penh as a high school student."
"Then, in Phnom Penh, I joined the Pol Pot regime."
"I had to go for my life and for farming in the Tohoku region."
"After that, I spent three years and eight months there."
"I spent a lot of work there."
"By chance, I was able to attend a local nursing school from the beginning of my20s to the age of25."
"So I was able to learn as a midwife and nurse."
"Then I started working as a nurse and midwife."
afterwards
"More than20 years have passed."
"Today, I continue to give lectures by midwives and nurses in each region."
"The lecture area is a major region in Japan, including the capital Phnom Penh."
"Mainly, I teach students between the ages of20 and 23 at school."
"One year is a two-semester system."

52 MEDICAL EXAM. CAMBODIA

"I'm in the lead in teaching from first to third grade."
The number of students who teach is like this.
"That would be hundreds"
This is a conversation with a busy nurse and midwife teacher who runs through Cambodia.

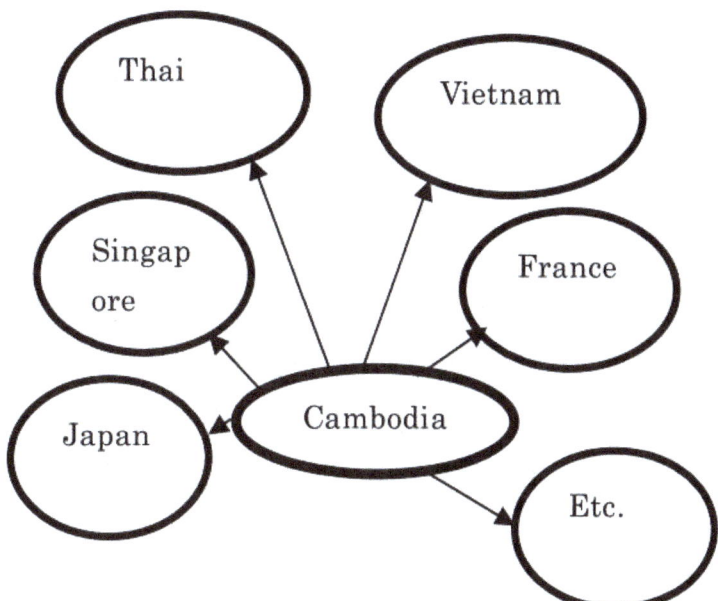

Student studying ABROAD's frame to approximately 7 countries per 60 persons to 70 persons.

54 MEDICAL EXAM. CAMBODIA

From previous page

※ In Cambodia, there are many people who think that many people will continue learning in many courses in the frame of a scholarship system.

It is a matter of course.

Passing in that about the frame of the medical science course who can't advance very much economically when there isn't the leeway seems like a most challenging frame, too.

Chapter 4

Female Doctors studying abroad under the scholarship Program

56 MEDICAL EXAM. CAMBODIA

Female doctor from the foreign countries?

From previous page

In Cambodia, the popularity centers on a scholarship system absolutely.
Even if it is related to the medical treatment, it is the same.
The woman doctor of whom a pass was accomplished excellently with the examination of a scholarship system and who returned from the foreign countries told the various things.
It was a valuable story.

58 MEDICAL EXAM. CAMBODIA

A female doctor studying abroad under the scholarship system

The new person is the person who is a town doctor in a local city today.
This person is a person who studied in Vietnam for medical science under the scholarship system of the country which wants to become a doctor when it is a high school student before.
Then again to Cambodia, this person is a woman to return home.
I was able to ask the woman who was a doctor by chance about the process of becoming a doctor.
To tell the truth, that was not coming from of the medical school in the national in Cambodia.
"I spent my late teens in Phnom Penh."
"I spent my high school in Phnom Penh."
After graduating from high school, this person passed the medical scholarship system between Cambodia and Vietnam. Under that, this person is a person who went to study abroad at the medical school in Hanoi, Vietnam.
The study abroad destination is like this.

"I studied in Hanoi, Vietnam"
"When I studied abroad, a couple of years of experience in a hospital was essential to me."
I went to this university from the age of 20 to 27.
"I actually experienced as a doctor at a hospital in Hanoi from the age of 27 to 29."
"There are two medical colleges in Vietnam, one of which I studied abroad."
"At my medical school, I had a course in the Department of Medical, Dental and Pharmaceutical Nursing."
"I couldn't afford my house, so I couldn't go to medical school at The Medical University in Phnom Penh."
Then, it is this way.
"I wanted to get on the path of medicine somehow under the scholarship system."
"In Cambodia, there are unexpectedly many students who have studied abroad in neighboring countries through scholarships with other countries."
"This is because tuition is free."

60 MEDICAL EXAM. CAMBODIA

"For example, I'm entering a medical school in Cambodia."
Then, it is this way.
"Tuition for a one-year medical course will be US$1,500 per year at the Royal University of The National University."
"It costs about US$3,000 at a private medical school."
"I was able to pass the scholarship system in Vietnam and Cambodia by chance."
In general, it's like this.
"In Cambodia, between 60 and 70 people a year are eligible for scholarships."
"The international students are studying abroad in Thailand, France, China, Japan, etc."
"And VIETNAM is in it."
"Under the scholarship program established between these several countries, students study medicine at overseas universities."
"I was able to pass the scholarship in Vietnam."
The examination subject of the scholarship system is like this.

"Subjects were chemistry, physics, mathematics, biology, etc."

As a result, this person went to study abroad at the medical school in Hanoi.

"Even in Vietnam, there are only two medical colleges around Hanoi."

"Another medical school is Hanoi Medical University."

After graduating from high school in Phnom Penh, he wanted to use the scholarship system to become a medical school.

Therefore, this person passed the examination under the scholarship system with Vietnam after graduating from high school.

That's why I studied in Phnom Penh from the age of 18 to 20."

And that. In the case of study abroad, it is like this.

"First of all, I will study Vietnamese for a year."

"After that, I study medicine from the age of 21 to 27."

62 MEDICAL EXAM. CAMBODIA

"After graduation, students are required to practice for three years of intern."
So, I gained three years of experience in a hospital in Vietnam.
Then, when I was30, I went back to Cambodia."
After returning to Japan, it looks like this.
"From the family situation, I decided to live in the current local city."
"Therefore, I live in the present place".
The hospital is from 7 a.m. to 6 p.m.
"I've just opened a hospital, so there are fewer patients."
By the way, I have to recall hearing this person's story.
The first is the same course of Vietnamese medical school and the medical school in Cambodia.
In common with the two universities, there are four courses in the Department of Medical, Dental and Pharmaceutical Nursing.
As for medical study abroad, this person who went to Vietnam thought about how to use the scholarship system enthusiastically.

On top of that, this person will make an effort, wondering how to become a doctor. This person will persevere and overcome difficulties.
This person overcame the hurdles of the scholarship system and passed the exam brilliantly.

Husband thinking of overseas hospital for hospitalization

In the 1990s, I had heard this.
I had heard that Cambodian patients are hospitalized in neighboring countries in Vietnam and Thailand to receive medical care for the treatment of diseases.
And, first of all, the patient tries to receive the medical treatment as a patient in Cambodia.
Then, medical technology is often not enough in Cambodia.
Therefore, the patient might dare to think about hospitalization to the hospital in Thailand.
In fact, there are many Cambodians who want that.

64 MEDICAL EXAM. CAMBODIA

Then, the patient cannot help relying on the country such as Thailand and Vietnam.
As a result, patients in Cambodia try to rely on foreign countries for medical care.
This is not a medical study abroad but a patient side.
However, Cambodian people cannot help going to the neighboring country anyway about the medical treatment.
In the 2010s, I had met someone who was actually a Cambodian businessman.
This wife has heart disease.
Therefore, the person was lost whether to be hospitalized in Thailand or not.
The first child, the person's wife, was able to give birth.
However, when the second child is given birth, the wife of the person put a strain on the heart.
Therefore, it was said that a highly difficult operation was necessary for the person's wife.
Then, this person thought about hospitalization to foreign countries at one o'clock.

"In Cambodia, patients are often forced to go to a foreign country when they are hospitalized."
It is said.

I'm going to Singapore to take overseas exams for the scholarship system!

I had to meet the protector of the junior high school student in capital Phnom Penh.
I also heard about the scholarship system.
"I am thinking about the frame of scholarship students overseas in Cambodia."
"To do that, I'm going to Singapore for the exam."
This was the words of parents who tried to win the scholarship system.
Then, in Cambodia, students have to pass extremely difficult exams.
"So my child dares to take the exam overseas."
I heard that.
This was a state of the examination which had to overcome a considerable competitive magnification.

66 MEDICAL EXAM. CAMBODIA

"If you're lucky, my child can go on a scholarship."
"My cousin was able to pass this."
In Cambodia, like this family, there are many people who try to study under the scholarship system.
From the junior high school graduation stage, when I entered high school, I was challenged to a scholarship system overseas.
Passing the examination of this system reflected to me as if it was an important target of the home for a while.
It is a serious event whether the scholarship system can be acquired so much.
It was my moment to remember that.
This scholarship system is also a listening system after graduating from junior high school.
This seems to be a fairly advanced way to go on to high school or university.
Moreover, the place has changed, and I have heard it also in the university student.

Even at the same university, students with scholarship programs and students who do not interact.
It is not a medical school system.
However, 30% of the students in the general faculty were in the scholarship framework at one university in PHNOM PENH, the capital.
That's how much the scholarship system is used.
It is also a great way for students to go on to higher education.

Chapter 5

Preparatory Schools attended by active medical, dental and Pharmaceutical Sciences

State of a class by a cramming school town.

From previous page

In Cambodia, it is like Asia that an entrance examination is terrible, and a free school, cramming school, language laboratory and so on are boiling in capital PHNOM PENH.
However, it isn't only before an entrance examination.
The people who go to a medical college often go to a cramming school to win the pass of the special subject implemented every grade after the entrance.

A medical school, department of dentistry and student of the school of pharmaceutical sciences go to a cramming school after the entrance.

From previous page

※ In Cambodia, it sometimes goes to a cramming school after the college admission.
It is because it is difficult to pass the examination of the special subject done for each grade.
It isn't possible to take a failing grade in a subject with a high difficulty level.
It goes to a cramming school for that.

Medical college students concentrating in Phnom Penh

Phnom Penh has three medical and dental universities.
Here, I give an example of the magnification of the School of Dentistry.
For example, between 300 and 400 students take the entrance examination for 45 dental schools. What you mean
"Of the 300 to 400 people, 45 can pass."
And that. For example, if the examinee is 400, the magnification is 10 times.
Anyway, the magnification of both the school of medicine and the school of dentistry is very high today.
For this reason, there is a specialized cram school in Phnom Penh.
For example, the story of four first-year high school students I met at a prep school looks like this.
"I go to preparatory school in a row with junior high and high school."

74 MEDICAL EXAM. CAMBODIA

These four students are enrolled in a difficult junior high school.
You are financially blessed families.

Medical college students from all over the place go to preparatory school.

I meet the student of the national medical, dental, and pharmaceutical university by chance in the preparatory school.
All are medical college students.
The students are from rural areas such as Kampong Cham, Sihanoukville and Battambang.
In addition, other students will be from PHNOM PENH, the capital.
They have graduated from high schools in the capital and local areas, respectively.
This is in the student district where the preparatory school is densely packed.
Anyway, I was glad to be able to meet active university students by chance.

Here, there is a food shop in the store, and there is a dining room inside the building. This is also where people can relax.
Because of this, junior high school students, high school students, university students, active medical college students, dental college students, and pharmaceutical students are studying hard.
There are some people from the capital Phnom Penh here, and there are people from various provinces here and there, too.

The students come from 24 states of the whole country.

From previous page

※ In Cambodia, a student preparing for taking an examination comes today from the various prefectures in Phnom PENH to an exam.
It is similar in all departments.
However, a medical department system university exists only in capital Phnom PENH.
Therefore, for example, a haze comes to PURIYAVI that a border dispute with Thai broke out formerly.
When that student graduated from a university excellently and became a doctor, it returned to a parents' home and was saying wanting to play an active part as a doctor age to it.

78 MEDICAL EXAM. CAMBODIA

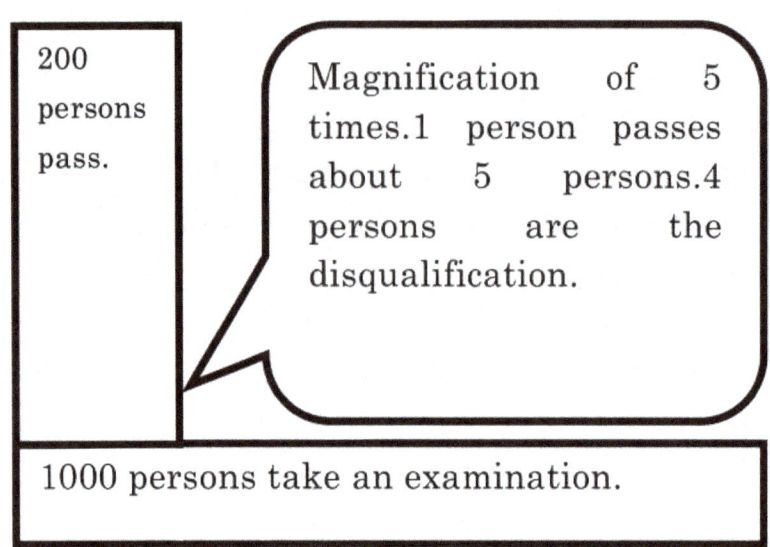

Barrier magnification that doesn't change both at the past and now

※ **The** magnification of a medical department system by Cambodia is about 5 times.

It is a quite great thing.

Because, in Cambodia, it is the body that isn't a system that all advances towards a grade above like Japan.

It is a method that a window is wide and an exit is narrow if saying generally.

80 MEDICAL EXAM. CAMBODIA

Medical University

This is the capital, Phnom Penh.
For example, this is an example in a medical university.
This is an example of the number of students recruited for each faculty.
The number of students in each faculty is as follows.
300 students in the School of Medicine course
Faculty of Dentistry course: 150 students
Faculty of Pharmaceutical Sciences course 100 people
Nurse course 90 people
Midwife course 90 people
A total of 690 people.
I hope that these people will continue to go smoothly in eight or five years. However, the reality does not seem to be so gentle.
Because, the examination of a special subject for each grade waits for the student.
This is the capacity of each faculty in the first grade.

Even after entering the university, I am studying difficult subjects at preparatory schools!

It's a conversation with a medical student
This means that you are learning such subjects, for example.
The subjects you study in the School of Medicine course will be like this.
Pathology/Pathology
Physiology/PTHOLOGY
SYMPTOMATICS / Semiology
Anatomy / Anatomy
Biochemistry/Biochemistry
These are examples of subjects studied in medical school. However, it becomes like this in the medical school.
"Students must finish10 courses."
"But the above subjects are particularly difficult."
So
"Even college students go to preparatory school."

82 MEDICAL EXAM. CAMBODIA

And explanation.

General medical physicians and general medical specialists

I meet the student who is enrolled in the medical school in Phnom Penh in the office of the preparatory school.
The students are financially blessed.
In addition, two students will join.
All three of the students met in the preparatory school.
In all, I was able to meet six students of the medical school.
I will hear various things from those people.
Three men at the medical school are like this.
"I'm aiming to be a specialized doctor/general medical specialist."
"In general, students can be 100% as general doctors and general physicians."
"But there's another hurdle for students to become specialized doctors."
For instance

"Trials are different for students to become areas of cancer treatment."
"So, 50%of the people who become doctors don't pass a single exam."
"Therefore, students will aim to be specialized doctors again."
"To that end, students gain experience and challenge the following year."
In a word, this is a reality that the graduate of the medical school is more serious.
Therefore, it is a current situation that a further national examination is waiting for the student.

*What is the difference between a general medical specialist and a general medical specialist?

The difference between a general medical specialist and a general medical specialist seems to have been caused by the difference in social structure and disease structure.

84 MEDICAL EXAM. CAMBODIA

The subdivision of advanced medical care requires even more necessary expertise.
Today, the shortage of doctors lies as it once was.
However, in addition to that, there seems to be a variety of medical sites in the medical world, for example, at the same time as the population explosion.
it is necessary to respond to each other and to enhance their expertise.
There are many patients in Cambodia. In addition, the therapeutic field has diversified, and expertise in detail is required. That is an urgent need for Cambodia, too.

Chapter 6

Situation of Foreign Languages in the FACULTY of Medicine, Pharmacy and Dentistry

A practicable language of a doctor by a hospital is shown.
For example, it becomes Khmer, English and French here.

From previous page

Because Cambodia was a French possession, it is strong in a French influence even in the department of the medical relation today, too.

It was decided to watch the pulse duty factor in an English and French university from a rating, necessity by each department.

88 MEDICAL EXAM. CAMBODIA

> French and English popular rate that doesn't change both the past and today.

| French 50% | English 50% |

About a department of dentistry, English has many.

A medical school separates as much as the fifty-fifty.

About the school of pharmaceutical sciences, are 80% the French mainstream?

From previous page

※ Because Cambodia was a French possession, a French force is strong.
It often comes across a French signboard here and there in a town.
A target has the side to say to be 20% for 80% on a learning surface, too, comparatively but in a medical school and school of pharmaceutical sciences, it still is a half-and-half situation.
It is 50% and 50%.

90 MEDICAL EXAM. CAMBODIA

All three pharmaceutical students happen to be university students in mainstream French.

I was able to talk to three students from the Faculty of Medical and Pharmaceutical Sciences at a preparatory school in Phnom Penh.
All three. In Cambodia, students enrolled in a medical dental and pharmaceutical university of a super difficult university.
"What do you learn as your first foreign language?"
I'm French.
"What is the second foreign language?"
I am in English.
That. It is something that I understand somehow. However, I decided to hear anyway.
This person is a person who experiences the way of learning a foreign language centering on French.
The first foreign language is French people.
All three are students from the Faculty of Pharmaceutical Sciences.
One is from SIEMREAP.

The second is from Battambang.
The other third is from Phnom Penh.

French at the pharmacy?

For me, it was in SIEM REAP in the 1990s.
It was when I stopped by chance to ask for medicine in the pharmacy next to my lodging house.
I was trying to use English for medicine.
At that time, the person who happened to be inside the pharmacy was a person who seemed to be the owner of the shop.
Then, it suddenly became like this.
Can you speak French?
And, on the contrary, I had been asked.
That moment is still remembered. However, it is such a way.
"French is of high importance in Cambodia."
And that's what made me realize.
Time has passed since then.
I am the one that the importance of French is still high to the student of the pharmacy

92 MEDICAL EXAM. CAMBODIA

department which spoke in capital Phnom Penh by chance.
Then, I am impressed, too.

Which foreign language do you like in each faculty?

I will ask the situation of the foreign language of the medicine pharmacy and dentistry system.
"What is the state of language choices in each faculty today?"
"In medicine and pharmaceutical sciences, foreign language learning is 50%to 50%."
"How about in the School of Dentistry?"
"French is 20% important in the School of Dentistry."
"English is 80%important."
In a word, it is like this.
"Some French was learned in high school."
"But in general, 20%of all high schools are French and 80%are English."
"In the Faculty of Medicine and the Faculty of Pharmaceutical Sciences, I have the opportunity

to learn French when I was a freshman in university."

Therefore

"In the faculty of medicine and the faculty of pharmacy, English and France are 50%each, so this is the percentage, which is average."

"For the School of Dentistry, English is more important, with 80%English and 20%in France."

"French is less important in the School of Dentistry."

And that. It's like this again.

"What percentage of the students who have passed university has experienced ronin?"

"About 10%of the total."

And that.

In the Faculty of Medicine and the Faculty of Pharmaceutical Sciences, English and French are50%popular. In the School of Dentistry, English is 80%popular and French is 20%.

Chapter 7

Nurses and midwifery schools born here and there

Delivery teacher

From previous page

It wasn't possible to grasp concretely about actually what kind of thing it is that a Cambodian population explodes. It is that it noticed, being told to nursing teacher, when moving Cambodia.
We didn't notice that nursing teacher, delivery teachers were the urgent business.

A transverse is a year, the vertical axis is person's numerical of a nursing teacher, delivery teacher.
In the place of an arrow, it increases.
It tends to increase from a little past 2010 of hit.
However, still there are few rates to a population.

Rapidly sought delivery teacher, nursing teacher

From previous page

※ In Cambodia, delivery teacher's necessity is a very natural thing.
However, we didn't notice at all that we did so at the 1990's first half.
It was the same in 2000, too.
The emergency of the person of a delivery teacher/nursing teacher thinks that it was immediately 2005 years later.

Born Nursing School

For me, it's in a hospital in a local city.
I hear about the nursing school more than the nurse in the hospital.
"Today, nurse's school is all over the country today."
"After graduating from high school, I will enter a nursing school."
I have finished my three-year curriculum.
"Then I have two years of interns."
"In a total of five years, I have qualified as a full-time nurse."

The number of students increasing in cities and rural areas

They are educated from around the age of 19 to around 22 and interns from around the age of 24. It is this way in the nursing school where it enrolled.
"There are about 1,000 students in total from 1st to 3rd grade."

100 MEDICAL EXAM. CAMBODIA

"There are about 300 students in each grade."
And explanation. Then, it is this way.
"There was an Associated Degree/Associate's Degree course."
"Under this system, I qualified for junior college graduation."
"So, I learned mid-term/midwife."
"I also had an intern."
"I finally completed and graduated from the course to become a midwife."
And explanation.

Another nurse's explanation.
I hear the nursing school in the hospital in the local city.
According to a nurse working in a hospital, it's like this.
I was at a local nursing school.
"The number and scale of the school's students will be like this."
"The school is a three-year system, and from 1st to 3rd graders is a regular study."

"After that, the 4th and 5th graders are interns."
"After all, it is the same as the previous one"
"It's a system called Associate Degree/Associate Degree."
"There are 330 first graders."
"There are 330 second-year students."
"There are 330 third-year students."
"I'm in the fourth and fifth grades, and I'm an intern."
"After completing those two years, I was able to obtain a nurse qualification."
And explanation.

Do you have an entrance examination to become a nurse? Missing?

"There are no entrance examinations at a vocational school called a nursing school."
"Of course, students don't have to pass Grade 12."
"Students should go to a vocational school with an associate degree system."

"Students only need to be motivated to learn."
"However, in the case of a college nursing department, it is different."
"However, grade 12 testing is important."
Students must pass the high school qualification exam.
"Students should have a passing score from A to E, which is the scope of passing."
"Students would like, for example, nurse and midwife courses at private universities."
"Then, students can advance to the desired university if they pass the pass score from A to E."
The qualifications of nurses and midwives can be obtained by students at vocational schools and universities.
It is not difficult to enter a vocational school.
In this case, the student may not pass the high school graduation qualification exam.

Bachelor's nurses graduating from university and nurses graduating from vocational schools

That is, in the case of nurses, students may not pass the Grade 12 national common exam.
First, students enroll in vocational schools with an Associate Degree/Associate's Degree system.
Students are required to complete specialized nursing and midwifery courses at vocational schools.
Students must pass each subject's exam for three years. This is where it matters.
Therefore, the student can acquire the qualification of the junior college graduation "Associate's degree".
At the same time, students can earn a nurse's qualification.
In addition, it becomes the medical school of the fourth grader here.
The nursing course here will be another course to become a nurse.
It's called a student, a bachelor nurse.
In other words, the student is not an Associate Degree/ "Associate's Degree". In other words,

students have passed the Grade 12 exam and went straight to the nursing department of the fourth-year medical school.
Because these people graduate from the university, "Bachelor's degree" is basically acquired at the same time as graduation. Students can earn the qualification of "Bachelor Nurse" once they have completed the internship.
In any case, there is a shortage of nurses in Cambodia, just like doctors and dentists.
By the way, the national examination for nurses begins in the early 2010s, along with doctors and dentists. Therefore, mature doctors and dentist nurses are those who have been employed in their respective professions at the same time as graduation without national examinations. Through the national examinations that began in the early 2010s, it seems to be a mix with those who became nurses.
However, in the case of mature doctors, there was no national examination by the early 2010s. However, the examination of each

specialized subject for each grade while attending school is very difficult, and it is serious until it arrives at graduation.

106 MEDICAL EXAM. CAMBODIA

Internship of 2 years — It takes 5 years.

3 years curriculum

Curriculum of a nursing teacher/delivery teacher

From previous page

※ Both the water and electricity were terrible even when Cambodia was the 1990's first half by a civil war.
Even if there was the water, the water in a water supply was the auburn.
It sometimes lit a candle and lanthanum at night.
The water, too, was the same as electricity or anything thing that isn't there.
It was troubled by a food, too.
We felt that we itched about what the rice that it has just cooked is.
However, there are all one hello.
It increases a population, too.
On the other hand, various patient-sans, too, appear.
Nursing teacher/midwife becomes the urgent business.
It is like this.

Chapter 8

Break time

Compare it with medical care in Cambodia, Thailand, Zambia and Japan.

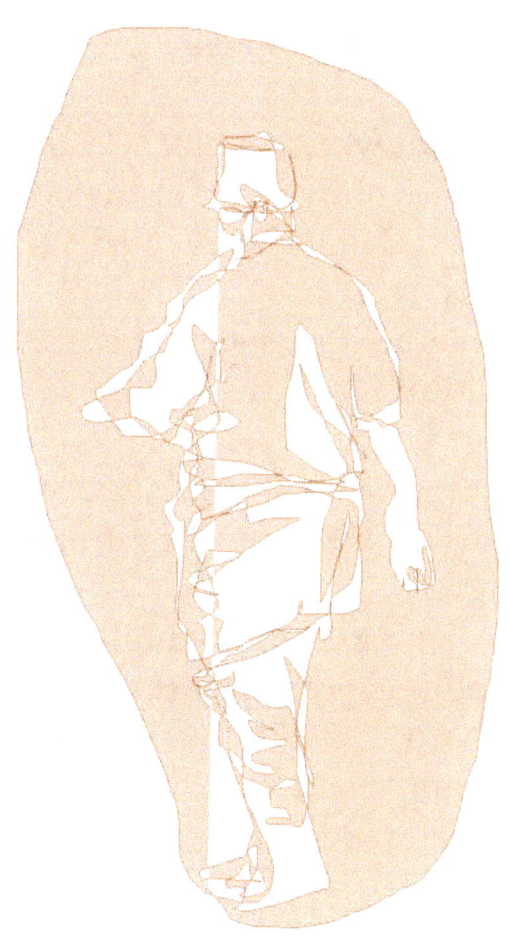

Doctor

From previous page

It not had ever thought of a thing such as the rate of increase of the occupation of the Cambodian medical world formerly.
However, hello, it seems that it could not help thinking of that thing in country itself.
A country is little and moreover isn't known like this but does a progressing feeling, too.
A duck isn't known gradually but it does the feeling more eased than today, too.
Also, we tried to compare with Cambodia, Thai, Zambia, Japan and so on in each field here.

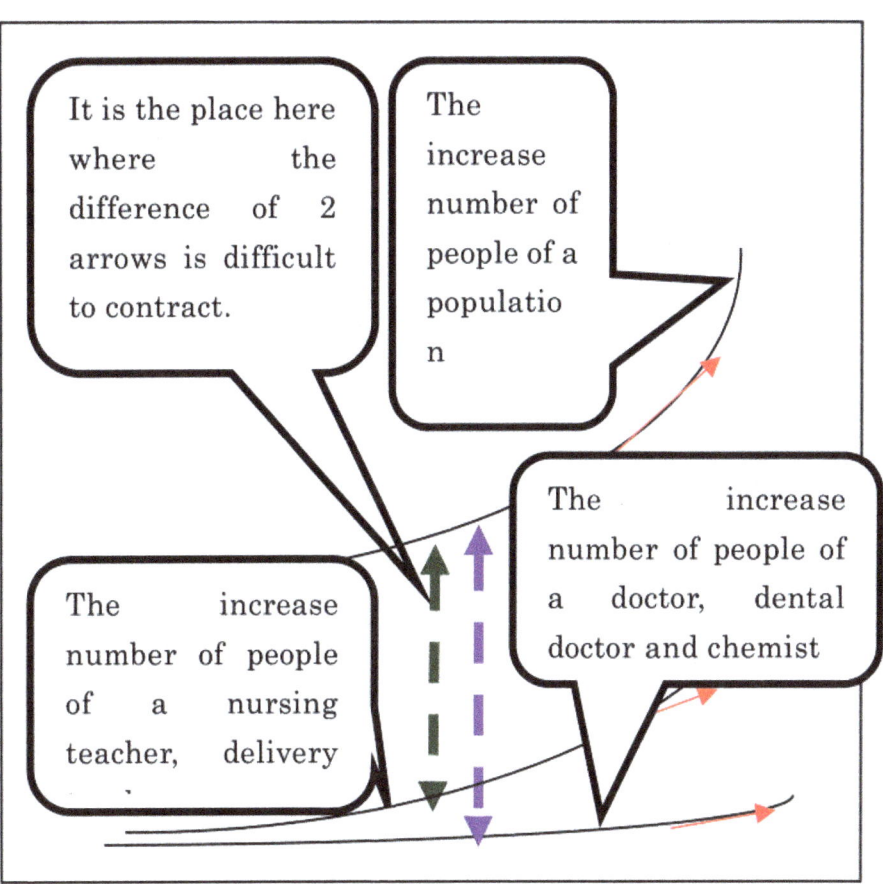

Number of the doctors who look up gradually on a figure but whom it is difficult for to quite increase.

From previous page

※ In Cambodia, it is a population explosion.
On the other hand, a matter of course, doctor and dental doctor can't do chemist's number about exploding steeply.
The rate of increase of a population and doctor's rate of increase are because it isn't proportional.

113

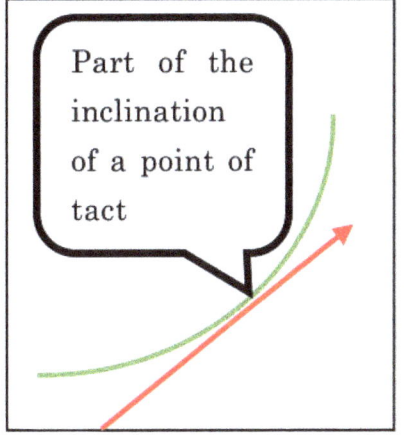

- An arrow is the rate of increase of a population.
Anyway, the rate of increase of a population is the highest.
The inclination of an arrow is the biggest.

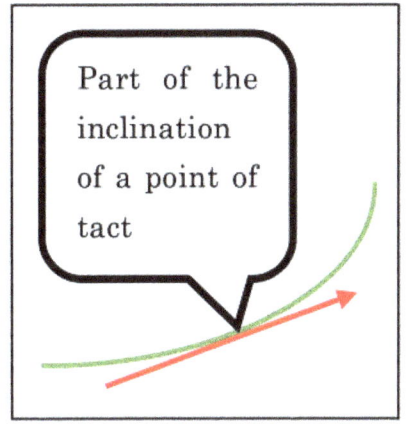

- An arrow is the rate of increase of a nursing teacher/delivery teacher.
It isn't as much as a doctor/dental doctor but is still lower than the rate of increase of a population.
The inclination of an arrow is a medium degree.

114 MEDICAL EXAM. CAMBODIA

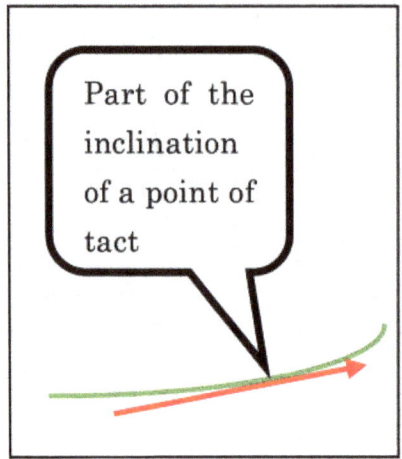

- An arrow is the rate of increase of a doctor, dental doctor.

Because the number of the universities is limited, it isn't in time for the rate of increase of a population.

The inclination of an arrow is the smallest.

In the early 1990s, around50 people, and then 8 years before graduation

A university specializing in medical science was established, and each faculty started. That was in the 1960s.
It is said that the number of doctors left at the time of the civil war is less than50.
Therefore, in the early 1990s, entrance examinations for each faculty became active.
For example, students entered the medical school in 1995.
For the student, it takes 6 years of study and 2 years of interning period to graduate.
This is 2003. This is the case with someone who was able to graduate well.
It is a tentative figure, but every year, about80%of medical students managed to get to graduate.
At that time, the number of students enrolled in the School of Medicine was 200.
About 160 of them will graduate.

116 MEDICAL EXAM. CAMBODIA

Ten years later, in 2013, this will finally become 1,600 people. At that time, the population of Cambodia is 16 million.

Even if the new university starts, it is still hard

By the way
Both universities were established in 1999 and 2003.
At each university, for example, around 2010, the university will finally send its graduates to the world.
It is 10 years later that this university can send doctors from the school of medicine, which has just started, into the world.

National examination begins

However, the university, the student, and even the graduation are serious. In addition to that, the national examination of the doctor will begin in Cambodia.

Until then, just graduating from medical college allowed students to go on to become doctors.
However, the national examination starts anyway in Cambodia.
In addition, the chance of the person who can leave as a doctor will be squeezed.
For example, I take it, for example, 2005.
The person who graduated from the new university will be eight years later.
At this time, the number of doctors for two schools certainly increased.
However, judging from the overall number of Cambodia's population, the number is small.
The national examination will also start.
Here, the medical college student who is forced to step comes out.

What about pharmacists, nurses and midwives?

The rate of increase may increase slightly, but nurses in a high position.
Many nursing schools have been able to be completed in Cambodia.

118 MEDICAL EXAM. CAMBODIA

Here, many nurses are born in Cambodia.
Short study and intern periods. Therefore, the number of people is in the increase rate that it is easy to increase compared with the increase rate of the doctor.
On the other hand, a pharmacist and a bachelor nurse who graduated from university are different from nurses.
Bachelor nurses have a little longer learning and intern periods than nurses.
However, it becomes an organization in the middle of shortening from the doctor.
The rate of increase will be between the rate of increase in nurses and the rate of increase in doctors and dentists.
However, this is much smaller than the rate of increase of the population.
So they are missing.

I compare it with some of the countries in the world.

This can be said the same thing also in Zambia in Africa.
There is only one medical university in the country here.
However, it is a country where there was no civil war here.
Therefore, doctors have been steadily produced in the world in the 1960s and 1970s.
However, there is only one medical university.
Therefore, the number of doctors does not increase suddenly.
The number of students in the School of Medicine cannot be expanded here.
In addition, the scene changes a little.
However, the social security system has not been established here. I don't have insurance in the hospital.
Therefore, people are expensive to treat.
On the other hand, the number of medical universities is far more for neighboring Thailand and Japan, etc.
So, the situation is much different between rich and non-rich countries.

120 MEDICAL EXAM. CAMBODIA

Moreover, the medical treatment facilities are completely different originally in such a country.

> It is a rapidly rising examination magnification medical school, department of dentistry.

Duple 40 in the duple in 30 in a magnification at the first half of 1990.

Duple 5 in about 2020 years of magnifications.

From previous page

※ The magnification of a 1990's department of dentistry by Cambodia was 30 times and 40 times.
It was the same a magnification in the adoption in a however outside hotel, too.
It is because there were too much few hotels on the side to adopt.
The number of the people who return home had a few places that work even if it increases too much.
Dental doctor's world, too, resembled.
Even if there are many persons who want to become, the number of people of call for the student of the dental college that accepts it is decided.
Increasing suddenly is because it isn't made.

Chapter 9

The entrance examination ratio of the FACULTY of Medicine, Dentistry and

Pharmaceutical Sciences

124 MEDICAL EXAM. CAMBODIA

Dental doctor

MEDICAL EXAM. CAMBODIA

From previous page

The thing short of a hand occurs in Cambodia simultaneously with a population explosion in the various fields.

It is short of a hand in 2008 in a medical field, too, in each field.

It was decided to reduce an entrance examination subject at last.

It was 2008.However, because an examination became easy, there isn't at all.

Because, it is because an examination continues.

It is a thing as much as being beyond imagination that advances towards doctor's road in Cambodia.

It is little even to this much and gets in a mood whether or not it may be squeezed.

Even if it does to be sometimes similar overseas, too, the competition in the situation like Cambodia exists in case of Cambodia.

The entrance examination ratio of the Faculty of Medicine, Dentistry and Pharmacy continues to this day

I visit the medical university.
It is one of the best universities in Cambodia.
I was able to talk to two students in the School of Dentistry.
Now, 24 years ago, when I spoke at a medical and dental university, I realized that there was a subtle change in the content.
Because, students that I met again were people who were from a local city, and had overcome the examination magnification of the difficulty.
This succeeded the story of students who belonged to the medical school of the university where I talked before.
That's because the probability that students pass the exam is about 5 to 10 times today.
This is not as much as the magnification before.
However, the process until the student arrives up to here is serious.

128 MEDICAL EXAM. CAMBODIA

Today, there is a serious process before students can pass the medical and dental examination and enter the school. Students must overcome the path of junior high and high school exams and overcome the competition before they reach the exam.

What if I fail?

However, I met the student of several dental schools afterwards. Then, the problem which does not understand newly has arisen to me.
Until 2008, it was difficult for active examinees and students to pass a difficult university.
At that time, the student who was not able to pass the examination had occurred.
And, there is no existence of a corporate preparatory school in Cambodia.
However, there are preparatory schools and supplementary school schools where students gather privately here and there.

First of all, in order to advance to a difficult undergraduate school or a difficult university, the examinee needs a certificate of completion of the grade 12 level in the third year of high school.
This is a Japanese high school graduation certificate. However, it is unexpectedly difficult for the student to obtain this certificate. This is a national unified high school graduation exam.
So, students win this Grade 12 level, and then take the medical college exams around November or December.
Therefore, this examinee aims at the national medical school for instance.
For example, more than 1,000 students gather in the number of successful applicants.
Therefore, 800 people failed.
The 800 students will take the entrance examinations at one of the few private medical universities remaining.
There are about 100 students each in the school of medicine and the school of dentistry.

130 MEDICAL EXAM. CAMBODIA

Then, the student takes the examination.
In addition, 600 students will surely fail here.
These 600 people can take the exam the following year after another year.
However, in Cambodia, the student will actually change direction.
Therefore, the student might be inclined to give up the medical school examination.

Do you want to retake the exam?

"How many medical and dental universities are there in Cambodia?"
"That's three schools."
How many students does the university recruit every year?
"It's a recruitment of 200 people each in each faculty."
"There are 200 students in the School of Medicine and 125 in the School of Dentistry."
For example, it's like this.

About 1,000 students take the entrance examination in the frame of the School of Dentistry.
So, the magnification of the School of Dentistry is around 10 times.
However, the admission quota is not necessarily 125 people.
In some cases, it is 90 people.
Also, it is 100 people. It may also be more than that.
Because, it is like this.
"Each university has a minimum pass score."
"If you don't go beyond that every year, your students won't be able to pass."
"The number of applicants is not determined at the university every year."
"The minimum passing score is the standard, and based on that, the university will give the number of successful applicants."
And the explanation.
Do students wait a year or two if they miss the exam?
"Students can wait."

132 MEDICAL EXAM. CAMBODIA

"There are no corporate preparatory schools in Cambodia."
"But there is a cram school like a specialized cram school in Phnom Penh."
"Going there, the students will take the exam again the following year."
"There are few students who take the exam for years, so many people change their course along the way."
"Who is the teacher of the specialized cram school?"
"I'm mainly run by high school teachers and cram school teachers."
"What kind of subjects are there for the examination?"
"There are many subjects about difficult faculties such as medical schools."
"It's math, physics, chemistry, biology, etc."
Is there insufficient number of doctors?
"Or is that okay with the current number?"
"Doctors and dentists are still short."
"Pharmacists aren't as good as doctors and dentists, but that's still lacking."

And explanation.

Due to the lack of doctors, there were also few magnifications and examination subjects.

Now, to become a doctor, students will need eight years.
Like doctors, it's been eight years for dentists. Pharmacists are required for 5 years.
This is the difference between the 1990s and today.
In the 1990s, the magnification was very high, about 20 times.
The time of the doctor examination was an examination in the school that there was no room for the student.
Today, however, the time of the examination became the examination in November.
It used to be around June or July.
If a student fails, the student must go to preparatory school.

In other words, students go to private specialized cram schools, such as chemistry, mathematics, biology, physics, and English. Students study subjects that they are not good at.
Then, the student will go to the cram school of the extra class which does the special lecture class.
Then I compare the subjects in the 1990s and today.
It is also a big difference that the examination subjects in medicine, dentistry, pharmacy, etc. have changed today.
In the past, it was an examination subject of about seven subjects.
Today, however, it is decreasing by four subjects.
Today, the essential subjects are mathematics, physics, chemistry, biology, etc.

Medical applicants change course?

But even so

"If you are an active student and you are aiming for a medical school or a school of dentistry, what to do if the student fails to take the examination?"

"Students will continue to challenge the following year, but if that's not possible, they'll change their course and enter another university or faculty."

However, it becomes like this.

"The number of students waiting for the next year is a minority.

The student changes his exams to almost other universities."

Of course, it's like this.

"Some students have never given up."

So

"From the 1990s to the 2000s, there were many students who were ronin."

Thus, the examination environment of the medical examination student in Cambodia has changed little by little.

It was remarkable that the test date was shifted in 2008.

136 MEDICAL EXAM. CAMBODIA

After the Grade 12 exam, students can study for four months for the next university exam.
As a rule, in Cambodia, the date of the examination system and the subjects for examinations have been changed and improved.
It seems that the number of lives has decreased more than more than 10 years ago.
Here, the shortage of the doctor was not solved at once.
Even if the number of examination subjects decreases, it is still necessary to enter the medical university.
Students also need a lot of perseverance before they graduate from college and clear national exams.
Across the sea, there is a similar situation for students in Japan and Europe.
In Cambodia, there is a violent landscape full of enthusiasm of medical exam.

Chapter 10

In 1994, the test magnification of the SCHOOL of Medicine was even more accelerated

138 MEDICAL EXAM. CAMBODIA

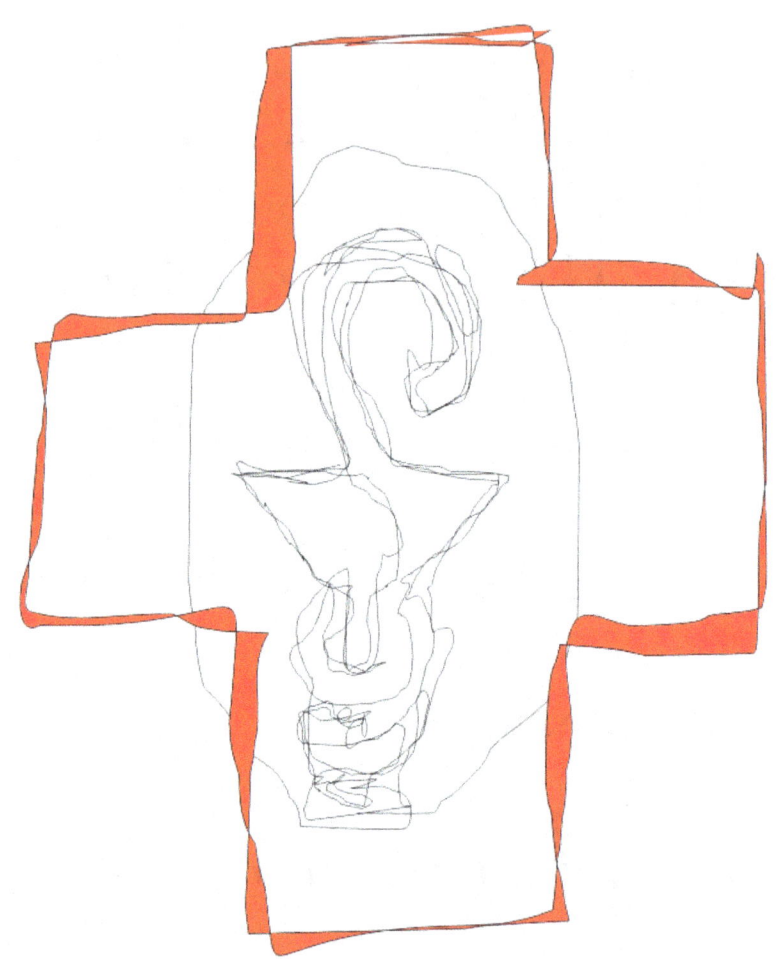

Mark of a pharmacy

139

From previous page

The trigger to make this pamphlet with still is because it had the opportunity to come across a medical school exam and one of a department of dentistry by chance in fiscal year 1994.

Then, a quarter of century passed but is still an order that a former magnification is remembered.

It decided to try to change there and to reappear in this pamphlet.

Preparatory School Status of University Entrance Examinations in General Faculties

For instance, this is in the preparatory school in the royal university.
There was a preparatory school adjacent to the site of the university.
This school building was for preparatory students.
In a day, students change in the building three times.
Therefore, this preparatory school was a system of "Replacement system".
Preparatory students were separated into three courses in the morning, noon, and afternoon.
Some of the students who study here are from rural areas. They are like this.
"I often live in a relative's house or in a pagoda's lodgings."
That. The number of preparatory students in this Phnom Penh university is like this.
"There are over 5,000 people here."

142 MEDICAL EXAM. CAMBODIA

There are about 80 preparatory students in the classrooms that have been granted permission. At the end of the class, the collection of money began.
The teacher collects 300 riels per person for a special lecture on mathematical trigonometric functions. Each preparatory school student pays money. In addition to this special lecture, "Normal one-year tuition will be 20000 Riel" And that.

University entrance examination status during the civil war

The competition rate of universities is about five times that of each faculty of general faculties.
The faculty of letters and the Faculty of Pharmaceutical Sciences are about 10 times higher.
I'm impressed by the fact that in the 20-year civil war, I've lived these prep school lives and lived intense preparatory school life.

The School of Medicine and Dentistry, surprisingly, abnormal. In the past few years, the competitive ratio of medical schools at universities in the Faculty of Medicine, Dentistry and Pharmaceutical Sciences has been around 7,000 examinees, compared to 50 successful applicants.
It is calculated as a magnification of 120 times.

University entrance examination for dental students

　One medical college student from a local city said this to me.
"In 1990, when I entered the school, it was30 to 40 times the magnification.
But now that the magnification has suddenly increased."
It was.
The student is currently learning as an intern.

144 MEDICAL EXAM. CAMBODIA

So, I meet students who are studying in medical facilities of the school's school of dentistry. This is how he goes to his school of dentistry.

"Here, there are about 200 students in all grades."

Dental students are also students who have overcome intense competition in university entrance examinations.

The person who is now in the fifth grade told me the magnification when I entered the school.

"Of the just under 1,000 applicants, 25 passed."

This is 40 times the magnification.

There are a lot of students who aim at the university while being in the midst of the civil war here.

Even if students do not find a place to find a job, there are preparatory students and university students who are aiming for university even though it is a very difficult employment situation.

That's here in Phnom Penh, and I've come to understand.

Because of the civil war, it may also be a way to aim for university.
"It's hot here in Phnom Penh! There was a lot of competition for exams."
That's why. There is a class for the examinee also in the school building of the university of the medical, dental, and pharmaceutical system. It is applied to it between about 11:00 a.m. and 2:00 p.m.
"The number of students enrolled there will be more than 1,000 people"
And explanation.

It starts from the advancement examination of the elementary school, and the entrance examination hell of the university is terrific, and it is Phnom Penh which comes into view to me.

The family structure is suddenly changing in the family which returns from the provinces. The number of people of one family becomes it. As Cambodia's population grows, so does the age group of adolescents.

That might also be in increasing the number of examinees for the entrance examination.
I was not able to catch the mechanism of the elementary school which was in front of this Phnom Penh easily.
However, I am made to learn whether the radicality of the university entrance examination is so terrible starting from that.
This is still the reason why the education of the reality inside Phnom Penh is still.
It's about civil war.
About refugees.
It's about politics.
It's about volunteering.
I am often informed about Cambodia and overseas help etc.
Such a thing is Cambodia with a lot.
However, the present state of the education and the examination competition were only unexpected for me like this.

Examination system of a French BACCALAUREATE

VS.

Examination system of Cambodian grade 12

From previous page

※ There are the various examination systems in the world.

If there isn't an examination both in an entrance and in an exit, any person has an equal feeling, and it may be 1st ideal system.

However, there isn't such one.

If an entrance is easy, an exit is severe.

If an entrance is severe oppositely, an exit seems to be easy surprisingly.

However, it has a feeling in case of a Cambodian medical department system university for either to be severe.

Or, it isn't known about whether or not to be "an entrance is more severe as much as a half than France", "in an exit, it is easier than France as much as a half", either.

Chapter 11

It is hard to enter and graduate from medical college, and it is hard again after entering medical school

150 MEDICAL EXAM. CAMBODIA

From previous page

It resembles the educational environment in Asia an appearance in the purpose in a system in 3 in 6.3 in a Cambodian education system above.
However, the contents are like Europe.
Student's number is shaken off every grade 10% at a time.
This seems to resemble a French BACCALAUREAT examination.

152 MEDICAL EXAM. CAMBODIA

It's hard again after I entered medical school.

Not everyone can advance to elementary, junior high, and high school in France when they go up to the first grade.
In France, first graders graduate from sixth grade, and between10%and 15%will have a year's stay.
When graduating from junior high school, it is said that the number of re-year-end groups will be around 20%.
At the university, you must pass the Baccalaureate Test, a national common examination conducted in France.
It is said that 80%will pass, but it is still difficult.
If you pass this, it is a system that anyone can enter the faculty of the school of medicine.
Therefore, there are a lot of people who hope to enter a school, and enter a school.
However, to tell the truth, there is a peculiar constitution in the medical school.

Every grade, students who are relentlessly unable to pass the exam come out.
As a doctor, only 10percent can go up to the upper grades.
Therefore, there are a lot of people who are forced to change course on the way, too.
The national common examination is the same for Cambodia.
Therefore, first, students must pass the Grade 12 National Common Test.
This is also the same national test as the French Baccalaureate.
In Cambodia, there is an entrance examination for admission of the medical university original again.
This is different from French medical college.
However, after entering medical school, you must pass specialized subjects every year.
This place is similar to France.
To put it simply, this is often said.
"Western universities are easy for students to enter and hard to get out of."

154 MEDICAL EXAM. CAMBODIA

"Japanese universities are difficult for students to enter and are easy to get out of."
It is said.
This makes such a difference.
However, the university in Cambodia seems to be difficult for me to enter and to go out.
This is what it is like.
Moreover, the cost is large by the time the student graduates if economically cannot afford it.

Is Cambodia a European way of progress?

Cambodia's education system looks Asian in terms of compulsory education.
The fact that it is the same as Japan is 6, 3.3 system.
The number of years of elementary, junior high, and high school is the same as the countries of East Asia and Southeast Asia such as China, Thailand, and South Korea.
However, when contents are seen, a European advancement method is still carried out.

The reason is that there are unexpectedly a lot of years as if it is a boyhood of the elementary school student and the junior high school student to repeat.

In France, compulsory education is 5,4.3 system. Obviously, it is not today's 6,3,3 system of Cambodia.

However, it is a system that the student who re-yearly often comes out for France and Cambodia in the middle of this system, even in the elementary school student, the junior high school student, and the high school student.

At the same time, in the final stage, there is a national unified completion examination.

Therefore, whether to pass or not is an extremely important signpost.

Anyway, in Cambodia and France, whether or not they can advance in each grade is an important issue.

If you fail, your students won't be able to go up to the upper grades.

If you can't pass, you have to stay for life.

156 MEDICAL EXAM. CAMBODIA

In France, there are exams again in the School of Medicine and the Faculty of Law.

Cambodian university entrance exams basically must pass the grade 12 high school graduation exam.
Otherwise, you will not be able to obtain a certificate of eligibility for graduating from high school.
This pass rate is50% over a span of about10years.
The annual pass rate can be around40 percent, or60 percent.
Only those who pass this Grade 12 are eligible for further entrance exams at the Medical University.
Therefore, those who have passed through the 5 to 10 times the frame can finally pass the undergraduate course of the Faculty of Medicine, the Faculty of Dentistry, and the Faculty of Pharmaceutical Sciences.

Do medical students go to preparatory schools in France and Cambodia?

France in Cambodia is also hard to graduate from entering a university.
France can still enter the school of medicine if it passes the national common examination.
However, it is difficult to pass through each grade after entering the School of Medicine.
In each grade, more than10%of people fail.
Therefore, about 10%of the students remain in the final grade.
Therefore, it goes to the preparatory school of the medical system to complete a specialized subject of the medicine though it is a university student while attending school.
In Cambodia, if you pass Grade 12, you must take an entrance exam for the School of Medicine and pass there.
However, after entering the School of Medicine, Cambodia still tries to digest the specialized subjects of medicine, dentistry, and pharmacy.

158 MEDICAL EXAM. CAMBODIA

Therefore, for difficult subjects, students go to preparatory schools.
Then, the student passes the examination.
Therefore, I become the atmosphere of going to the preparatory school until it enters the university because it is a medical school to the desire that both countries look like.
The French Baccalaureate exam is known as even more difficult, especially for the Grande Cole exam in the Faculty of Law.
However, there is a more difficult examination about Cambodia as the competition rate of the medical school and the faculty of law, etc. is violent. For students, there are barriers ahead.

After all, are medical schools, law faculties, and popular faculties difficult?

In other words, from a comprehensive perspective, it is difficult for students until the school of medicine and the faculty of law graduate.
In addition, this is in other faculties.

In Cambodia, there are new exams in the faculties that are popular in medicine, law, and other faculties.
However, many universities can enter a university if they pass Grade 12.
Even if it is an examination of medical school, "C" and "E" which hits the lowest in the grade may be good.
It is not because it is a medical school that it is not had to be "A".
Still, there is an examination qualification.
In addition, other faculties do not rate universities.
In terms of grade 12 pass rates and so on, I can tell you the difference.
However, it is similar to Baccalaureate in France.
Even baccalaureates in France can enter the university if they pass this exam.

There is a unique medical examination environment in each country!

160 MEDICAL EXAM. CAMBODIA

By the way, I'm not just saying that the French Baccalaureate and the Cambodian Grade 12 system are similar.
There are examination environments in various countries.
The medical examination in any country cannot be overcome if there is no considerable perseverance and the desire to study in the examination.
Otherwise, students will not continue to learn.
In that sense, it might be common.
The economic power to continue learning is also necessary for the student.
Then, it is not the same as the medical examination in Japan at all.
However, I am not saying that here is the same as the medical system examination in France.
Cambodia might have been influenced in such a scene because it was a colony of France.
However, I might have wanted to know that there was still an examination environment where the medical examination of the Cambodian style was specialized in Cambodia.

In addition, this is different from the examination environment of tens of times the medical examination in the civil war age. Also, at the same time, I'm in a situation where I face a scene of a new exam competition.

Chapter 12 Rehabilitation welfare facilities

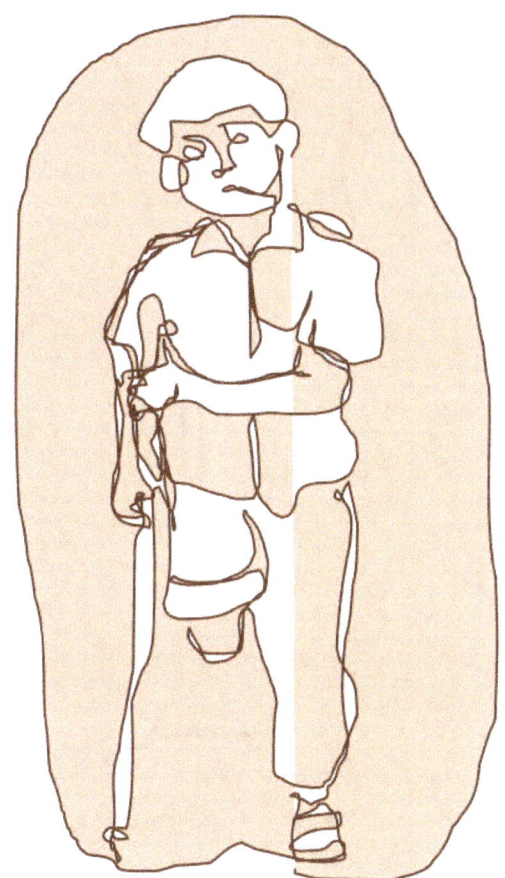

Disabled person

Artificial limb equipment

164 MEDICAL EXAM. CAMBODIA

Artificial leg

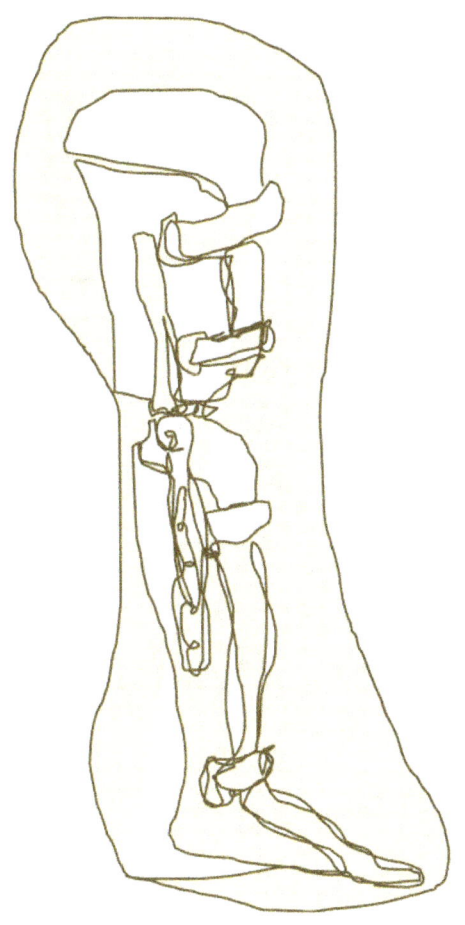

Considerably highly precise artificial leg

166 MEDICAL EXAM. CAMBODIA

Wheelchair

From previous page

It comes across the facilities about a rehabilitation in a local city.
Such a field, too, is the characteristic of one of the Cambodian medical worlds.
Usually, it may be the place where orthopedist's here and there and person of a physical-therapist are playing an active part if saying in the medicine.

168 MEDICAL EXAM. CAMBODIA

People living in wheelchairs

Originally, I only have to bring this booklet together in the condition of medical treatment and nursing care welfare.
However, I do not end still only by it. It becomes a medical facility peculiar to The Cambodia.
Because it is Cambodia, what cannot be overlooked by all means comes out to me.
This comes out because it is one of the world's leading landmine powers.
That's the building of the rehabilitation center.
I did not go directly into the building and visit it.
Just I was walking on the road.
Then, such a building happened in front of me.
This building was a rural building.
However, people with disabilities are forced to live in Cambodia.
In front of the building, there are patterns of people who have lost one leg or people who have

lost one arm. Then, it understands to me that
the building is such people' buildings.
Because Cambodia is a landmine country and it is
a mine that will never be easily lost, it is for
buildings that will continue for a long time.
Here, doctors who are surgeons and
physiotherapists are playing an active role.
There are people who continue recuperation so
that the body improves to some extent, and
recovers further afterwards.
This is a building with such patient people.
This is a building and facility unique to
Cambodia.
This is neither a hospital nor a nursing home in
a so-called rich country.
It may be a welfare facility, but the situation
in a rich country is different for me.
In such facilities, orthopedic surgeons and
physiotherapists may be a field that is greatly
involved.
This is because it is the facility of people who
were damaged by landmines, grenades, time bombs,
etc.

170 MEDICAL EXAM. CAMBODIA

If you become a specialist who supports these people, you will need the educational institution. This is the reality of Cambodia's medical care.

A variety of prosthetic braces, not just wheelchairs

I sometimes see people in wheelchairs at the border.
I see that the convenient one is used for people with the physically handicapped because there is an electric wheelchair etc. in the country where it is rich.
In a place that is not a rich country, I wonder how people with disabilities spend their daily lives and move their bodies.

I have been unconsciously holding such daily life, and acting in Cambodia.
At such times, I sometimes see it by chance.
It is such a way.

I see the saddle that the person's both hands are directly connected to the gear in front of the eyes.
The saddle follows the gear, and the gear follows the chain.
Then, the chain follows the wheel of the wheelchair.
It is the one that Cambodian people devised even to the level whether the rotation person can be devised so much.
For example, this is not an electric gear.
Even so, I see the appearance of people who are active daily by a manual wheelchair in the border.
This wheelchair makes square lumber boards a floorboard.
Then, the pipe is welded well, and the bicycle etc. were remodeled.
Therefore, it was really devised by the hand of the person and it was produced.
I become such feelings.
I have seen the one which looks like this in the vicinity and around Vietnam.

172 MEDICAL EXAM. CAMBODIA

However, I admire that this was really born from the wisdom of life.
However, it is not at all only by this.
The instruments depicted in this welfare building support the people living here.
It is generally called prosthetic braces.
It is a brace subtly devised in the connection part of the arm and the foot.
It reflects to me like an excellent brace thought out so that friction never comes out between the lost part of the foot and the arm so that the friction never comes out.

Epilogue

In the early 1990s, sick people were on a street that took a step from the boulevard.
Then, there were sick people in the building of one story named the hospital.
It was frequent that sick people lying on simple beds in the building happened to come into my sight just by walking down the street.
Sweating from my forehead, I'm walking.
Then, the appearance that the patient was waiting in the building often entered my view as the drip seemed to be given to the patient.
Such a appearance did not enter my view by chance in Phnom Penh today.
The building improved, and the facilities improved.
However, in reality, Cambodia's medical facilities are still not sufficient.
In Cambodia, it seems that it is still necessary to be enriched enough in various fields in respects of medical medicine, dentistry,

174 MEDICAL EXAM. CAMBODIA

pharmacy, nursing department, and midwifery department.

Meanwhile, the mind whether the thing that I saw was able to be connected further came out to me through the medical treatment examination a quarter of a century ago.

Originally, because it is medical treatment, it might be appropriate to write the site of the medical treatment and the problem, etc. Therefore, I decided to make the work by feelings of being able to use a partial use of the site in today's medical examination in Cambodia in this booklet.

Finally, if you have any questions in this booklet, please be overcome on your own.

Angkor Wat・Cambodia Books etc. From Senkawa Tomoo

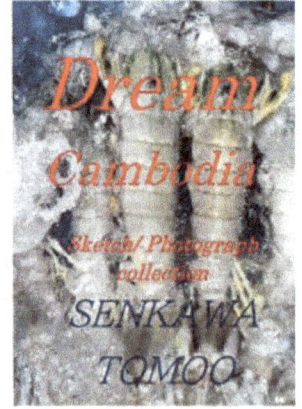

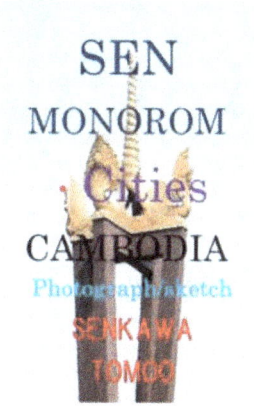

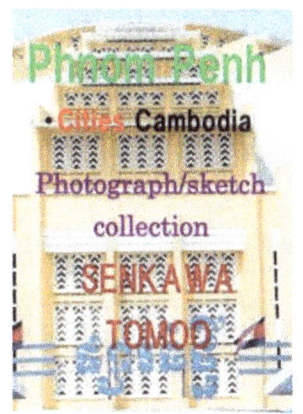

etc.

www.ingramcontent.com/pod-product-compliance
Lightning Source LLC
Chambersburg PA
CBHW071402210526
45465CB00001B/218